Narrow Gap World

Qiao He

Copyright © 2023 Qiao He

All rights reserved. No part of this book may be reproduced, translated, excerpted, reprinted, or adapted in any form or by any means without the written permission of the copyright owner.

Cover design: Asian Culture Press LLC
Cover illustration: Qiao He

ISBN: 978-1-957144-92-4
Retail price: US$13.00

Published and Distributed in December 2023 in Boulder, Colorado, the United States.

Please contact the Publisher for any copyright inquiries, information on reprints, adaptations, or other licensing inquiries.

Asian Culture Press LLC
1942 Broadway St., Suite 314c, Boulder, CO 80302, United States
Email: info@asianculture.press

This book is dedicated to ZUN, the shinto priest of TouHou Project, who has a wrong view of the gap world.

Preface

Human beings are newborns expelled from the gap world, with great potential and lack of understanding of the world. Their existence threatens the balance of the gap world, so they are "banished" to earth at a young age. However, there are children who have disappeared from the mortal world because it is difficult for them to survive on earth due to their abilities and future, and they have been "hidden" in the gap world.

Content

Gap World Theory 1

Introduction to Narrow Gap World 3

Narrow gap world (1) 5

Narrow gap world (2) 7

Narrow gap world (3) 10

Narrow gap world (4) 12

Intensive Talk on Narrow Gap World (1) 14

Conclusion 42

Annex 1 46

Annex 2 65

I. Gap World Theory

It's a new concept, there's basically no relevant literature or papers, and it's just fragmented pieces if anything, and it's like talking gap world but nothing relevant nonsense. So I'm going to describe the narrow gap world in simple and concise terms that can be understood by everyone.

First of all, what is the gap world?

The gap world is divided into a narrow gap world and a broad gap world. Let's put aside the broad gap world here.

II. Introduction to Narrow Gap World

The Narrow Gap World (hereinafter referred to as the Narrow Gap) is a place that is like a land of cockaigne, separated from the mortal world. But it's not as good as you think it is, because it's really just mortal world, with a few limitations added on top of it. For example, if y=f(x), y is the mortal world and x is the narrow gap world, and they have an extra f( ) to restrict each other, and that's how it works.

III. Narrow gap world (1)

The narrow gap world is dominated by one person, and only one person, while the broad gap world is a world dominated by more than one, a team or a group of people. That's the difference between the two kinds of gap worlds, and there are many differences, of course, so let's move on to the narrow gap world.

IV. Narrow gap world (2)

We were just talking about y=f(x), so what is f( )?

Let's do this first, leave f( ) alone for a moment, you need to understand what x, the narrow gap, is first.

Comparing the gap to a desktop computer that can't be connected to the Internet would be a more fitting image of the gap, and it would also be more appropriate. Never underestimate the narrow gap theory, it's not an unfounded pipe dream because even 5 and 6 year old kids are exploring their narrow gap these days. It's just that they don't know what they're exploring, and this is even more true of adults.

The computer is used as a metaphor for the gap because human development is the gradual materialization of the gap world.

V. Narrow gap world (3)

The narrow gap is not a single gap, it is made up of multiple gaps. In plain speak, one desktop computer is no longer enough for you, and you have a total of two desktop computers and two laptops kept in your house. All for you alone, of course. That's important. Oh, and by the way, there are two or three computers that can't connect to the Internet.

VI. Narrow gap world (4)

Okay, so now you know something about X, right? Now let's talk about f( ), the limitation. y=f(x), y is the mortal world, but 80% or 90% of it has basically nothing much to do with you, let's take out those parts that don't have much to do with you, and leave only the parts that are part of your home and life around you. Of course, the people you know, the people you work with, your neighbors, the people who live nearby that you don't know yet, all of these people are still there, and rest assured that it won't affect your normal life, shopping, exercising, working, playing, none of that will change. But if you want to go on a trip, you'll realize something's gone wrong, something's out of place.

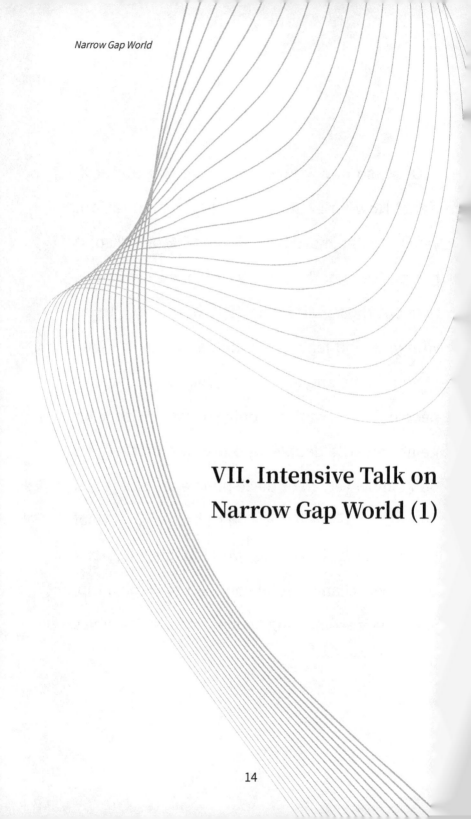

VII. Intensive Talk on Narrow Gap World (1)

The sun rises in the east and sets in the west, and the moon rises as usual; these are normal phenomena. But if you go out traveling by car on your own, you will find that you drive from the right side of your home and keep going in a straight line, and when you drive until almost noon, you will find your home appearing in front of you once again, and you come back again from the left side of your home.

What's going on? Is the Earth getting smaller?

Yes, the Earth got smaller and it's not called Earth anymore, it's called the Gap World.

In this gap, you are just mechanically repeating your daily chores that are basically the same, and although it's enjoyable to chat and drink with your friends, you still feel that something exciting and fresh is lacking.

By the way, you won't see news about the national government on TV anymore, nor will you see news about another conflict abroad. In this gap, not only space is limited, but time is equally limited. Time on earth is from zero to infinity. In the gap world, time is finite and constantly coming again and again. But when does it come back, it can be any number, 5 years, 50 years, 100 years? But preferably not more than 100 years, the gap is meaningless if you die.

Intensive Talk on Narrow Gap World (2)

The length of the cycle depends on the dominant person in the narrow gap, if the life of the dominant person on earth is too difficult, there are no easy days at all, and they are pressured to death by heavy official duties every day, and are full of confusion and anger inside (of course, I think that all youth workers nowadays should be in this kind of mentality, ah, including myself.) Such a person usually chooses a short cycle of years, often a few years when they have a better life, but of course the number of years will gradually increase when they have a family and a career, and the number of years in the cycle is not fixed, it can be changed. But every time you change the number of years in the cycle, the contents of the gap world will repeat what happened in the following cycle years, based on your current situation.

Intensive Talk on Narrow Gap World (3)

Of course, if you compare this gap to a desktop computer, you can use a USB flash drive or removable hard disk to make a good data backup, so that it won't be too late to repent.

For older people, the leisure time will be longer, and such people will of course want to turn in the cycles in which they are in good health.

For children, the gap is perhaps at its most powerful, which is why it is said that humans are children of great potential banished from the gap because they threaten the balance of the entire gap.

Intensive Talk on Narrow Gap World (4)

So what happens when the gap is given to children?

At this point, I didn't bother to think about it in depth because it's beyond my ability to think, and all I can say is that the gap will change.

As for what kind of change, don't ask me, I don't know.

Intensive Talk on Narrow Gap World (5)

Okay, we're done with the length of the cycle of turns, and then we'll talk about the other people who survive in the gap.

The person who inhabits this gap world, since the dominant person is only yourself and there is no second one, you can compare the person who inhabits the gap world to a computer person.

But don't underestimate these computer guys,

they are perfect replicas of what they used to be in reality, and they can reproduce normally. The only difference is that their memories will disappear. At the end of the cycle of turns you have enacted, time will rewind back to its initial position and their memories will disappear with it, just as if your computer had been reinstalled and each reinstallation emptied the contents of the computer. As for yourself, you are the operator, of course the memory is still there.

It should be corrected that the memories of the computer person do not "disappear", but "revert" to the memories they had at the beginning of the cycle.

Intensive Talk on Narrow Gap World (6)

Well? Do you understand everything I've said so far? If you can still keep up with what I'm saying, then congratulations on having realized the essence of the narrow gap theory. If you didn't follow it, review it by yourself a few times, and we'll move on.

Intensive Talk on Narrow Gap World (7)

Well, the first gap we have talked about, this gap is the most important part of narrow gap, let's take this desktop computer, that is, the gap, named it "Harmony".

Well? Why "Harmony"? You'll find out later.

The computer "Harmony" only contains the mechanized daily life, and once you start to get tired of it and crave for excitement and novelty, the computer starts to stop working.

And let's not forget that the computer "Harmony" is not connected to the Internet, and since time and space are fixed, they can't receive news from the outside world and can only communicate what's happening around them. The big events that are fixed to happen in cycle time will still happen, the only variables are only the people, and the development of the people.

Newborn children in cycle time may threaten the equilibrium of the gap world, while the development of people does not really have much of an impact on the gap world where time has been fixed.

Well, now that the basic variables have been analyzed, we can stop with this computer, but remember, the computer "Harmony" is the most important part of your entire narrow gap.

Intensive Talk on Narrow Gap World (8)

When you're craving freshness and excitement, the computer "Harmony" doesn't work. Let's turn it off and unplug it.

Think back to the question we started with. What is the narrow gap?

It is made by a dominant and contains a large gap made up of multiple gaps.

So, we're only talking about a desktop computer "Harmony" now, and there are actually several other computers inside the narrow gap, right?

So for the second computer, instead of a desktop, let's use a gaming laptop, isn't that fresh and exciting?

This second gaming laptop we named "Wild".

What's inside this computer is very simple, a single rule, a single model. Let me give you an example that is the easiest to understand: "Choose the character of your desire, pick up a weapon, keep practicing yourself, leveling up, keep challenging the stronger ones, keep winning, there's no room for failure here, and only by constantly forging ahead will you be able to survive in this place.

Isn't this example very easy to understand?

The "stronger" mentioned here can be a high level computer or a live action (networked game, that is), it depends on what kind of personality the dominant player has, some people like to challenge the high level computer, some people are eager to challenge the professional players, that's all.

This is the point at which you can tell if this computer is networked or not, depending largely on what the dominant person thinks.

By the way, the whole time the dominant person is not required to be physically involved, just manipulating the computer person. (That's playing the game.)

Okay, now here comes the somewhat more difficult, what I just said is the simplest example. "Wild" can take any form, the computer is meant to be a place for the dominant to vent their anger and make progress, but it is possible to dehumanize yourself, and if humanity is dehumanized too much, "wild" has no meaning.

Methods are actually important here, how to vent anger and how to make continuous progress, all of these need to be explored by the dominant person on his/her own, or he/she can ask for help from outsiders. (Of course, there are very few people who can give the right guidance now, because the concept of gap is not theorized and no one knows about it.

I hope that in the future, there will be a profession called "gap tutor"). In the world of "Wild", you need to find a way to vent your anger that really works for you, and keep it as short as possible. No dominant person should stay in the gap for too long, because it is dehumanizing.

Intensive Talk on Narrow Gap World (9)

Okay, so we're basically halfway done with the narrow gap, and there's not really much to the other half, but we're going to finish it.

The "Wild" is a gaming laptop, and then we pulled out a third computer, which is also a desktop computer, and unlike the "Harmony", this one can't store anything in it, and as soon as it's turned off, everything in it disappears. But it has one advantage, it connects to the Internet and allows you to discover new and interesting things. Let's name this desktop computer "Novelty".

Is this computer just for discovering new and exciting things? Yes, that's all it does. It's empty inside.

If you find something too interesting and want to keep it, get out a flash drive or portable hard drive and store it.

Intensive Talk on Narrow Gap World (10)

The last gap is called "Lucky". This time, let's just change to a windows computer with a gamepad. (It's important that it's a computerized gamepad.)

Doesn't the name make your heart race?

The computer "Lucky" contains the dominant's most prized possessions and the rage of "losing what they love", both of which are actually very dangerous and will give you a heart attack, don't touch the computer unless you have to or you will die.

The dominant person's most cherished possession and the anger of "losing what they love" are difficult to explain, depending on what the dominant person is thinking.

Narrow Gap Conclusion

Okay, this is our narrow gap, which consists of four computers, "Wild", "Harmony", "Lucky" and "Novelty". It can be more or less than that, depending on what the dominant person wants.

However, I think that the narrow gap formed by the four gaps of "Wild", "Harmony", "Lucky" and "Novelty" is the most ideal narrow gap.

The above is the full report I did on the narrow gap.

Introduction to Broad Gap World

What is the broad gap? I mentioned at the beginning that it is a world dominated by more than one, a team or group of people.

(The following is only my own speculation, and is only a brief discussion due to my limited mental capacity) It can consist of multiple gaps, or just one. It differs from the narrow gap in that there is no limit to the time in the broad gap, but it lags behind the mortal world. But since it's infinite in time, and would certainly keep up with reality if it were to develop normally, it would restrict the minds of the people inhabiting the gap world. I have a strong aversion to the whole restricting people's minds thing, and the fact that it's still jointly regulated by multiple dominants is beyond my ability to think, that's it, stop there.

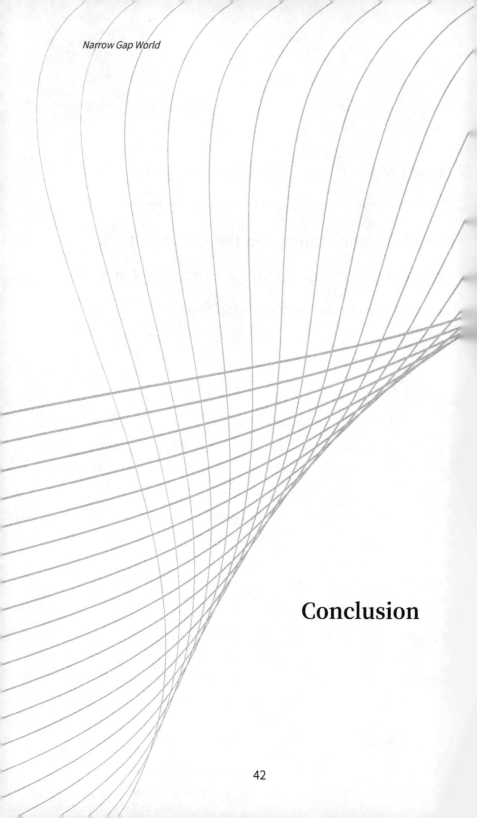

Conclusion

I've always explored what it would be like to be at the height of one's talents and abilities, and now I have a pretty good idea of what it would be like: weak, meaningless, and better to go through life happily every day. However, I have gained quite practical knowledge along the way of my ongoing quest, and if these realizations are conveyed to future generations and give them some insight before the discoverer dies, then perhaps my journey will not be pointless.

ZUN's Curiosities of Lotus Asia contains an excellent telling of the wide world of the gap, and I'm not going to complain about why the majority of the characters appearing are female, ZUN has his own take on that.

Lotus Land Story is a place that restricts people's minds, but ZUN portrays it in a very intimate and relatable way, so maybe that's an ability I can't fathom.

Probably the most fundamental difference between the narrow gap and the broad gap is also the inability to understand each other.

Narrow Gap World

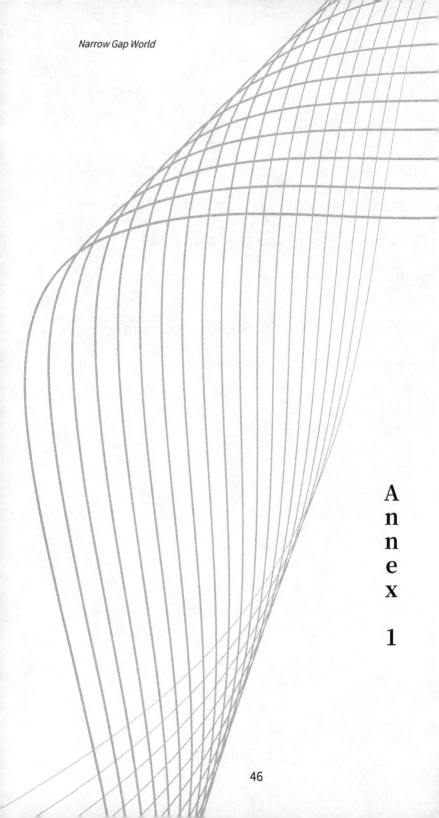

Annex 1

Notes on Curiosities of Lotus Asia

I always develop new insights while reading this book, I don't know why, it's like a class group discussion, when someone goes against you, it stimulates your mind, I've written these feelings in the annex.

Curiosities of Lotus Asia P102

Original: The gods originally had no names, ----- These gods with their now well-known names represent, at best, one of the many faces of God. ----- The fact that giving a name changes the nature of a god proves that a name is one of the many faces of a god. God's original appearance is actually much wider and indistinguishable from the nameless.

Conversely, God, who maintains the original look, can only be lodged within an object that has not yet acquired a name, because God, even when lodged within an object that already has a name, can only manifest one of the many faces of God.

Notes: I think the gods are pure, there are gods with large power and gods with small power, the gods with large power usually have names, the god of wind, the god of thunder. A god with small powers, so small that it doesn't have any power at all, it's just a very pure thing, but it's something that you respect very much. You try to find a name to describe it, but it's hard to find a corresponding word, because its power is so small, and you end up summarizing it with the word "god".

P139 Original: Matter is established as wood, fire, earth, gold, and water. When these five elements interact with the Ten Forces, they do not maintain stability and continue to change forever. The Ten Forces mean that wood produces fire, fire nurtures earth, earth nurtures gold, gold purifies water, and water nourishes wood. Wood can reduce earth, earth can absorb water, water can reduce fire, fire can melt gold, and gold can break wood. These ten complex interactions are not limited to two elements. When the power of gold weakens, the power of earth becomes stronger. But when wood becomes strong, fire becomes strong, and earth becomes strong, which in turn makes gold strong. When gold becomes strong, it weakens the earth, making it absolutely impossible to maintain stability and continually operate, causing various substances to be born and disappear.

Note: I have nothing to add to this passage. I was thinking of something else: the yin-yang diagram.

(Note: The following, believe it or not, is pure bullshit.)

My experience has taught me that yin and yang are not going to be balanced, and that the nature of the world is ultimately more black than white, and more evil than good, so it is possible that the yin and yang diagram is wrongly understood. Anyone who knows anything about wildlife generally realizes that it is always the vicious, brutal, evil-doing carnivores that dominate, and that herbivores are only hunted. Humans are animals too, and it's right to want to change that code, but I don't think it's appropriate to do so. Creatures that deviate from the fundamental guidelines are like herbivores and pay a terrible price.

P188 original text: Unlike demons, gods and spirits always have two personalities. ----- These two personalities are called "Harmony" and "Wild". Among them, "Harmony" is the personality that is kind to human beings. This is usually called divine virtue. ----- Isn't it said that the gods are the origin of all things? The personalities of the gods are the personalities of all things, so the feelings and power of the gods will be directly expressed in material things. ----- If further subdivided, "Harmony" can also be divided into "Lucky" and "Novelty". "Lucky" enriches the heart, and "Novelty" imparts knowledge. ----- "Wild" is the wrath of the gods, and the retribution is the manifested part.

It is the personality of the wild that is the true power of the gods. Part of the sacrifice to the "Wild" is related to the protection of mankind from the threat of the evil side, -----, and the miracle is to appease the personality of the "Wild" and to remember the personality of the "Harmony" in order to strengthen the power of the gods.

NOTE: Great, ZUN, we're in rare synchronicity. I call "Wild, Harmony, Lucky, Novelty" the four phases of the human heart. The meaning is similar to what he explains, but the only difference is "Lucky", which I think represents danger. What kind of danger, good or bad, is unknown. If you do it successfully, you are "fortunate". If you do it badly, it will be "disaster".

P201 Original Text: This world is established by a three-order structure. First, there is the physical order in which creatures and props act according to certain physical laws. The fact that objects fall towards the ground and rivers flow belongs to this order.

The second order is the inner movement and the mental order such as magic or sorcery. Meeting a nasty guy will affect your mood, and organizing a party to bridge the gap is this order. Most monsters can only grasp the world through two levels of reasoning, the physical and the psychological, which is why they make statements such as history repeats itself and the future is predetermined.

The third order of the world would seem to reject repetition. This third order is called the memory order in which everything remembers established facts. The memory order only increases, it does not decrease, and therefore it is impossible to present the same state as in the past. If it were truly identical to the past, it would leave the same memories that were transformed into a past with nowhere to hide, leading to a contradiction, so the memory order increases rather than decreases.

The physical order has the laws of physics, the mental order has the interpretation of the results, and the memory order manipulates the chances, forming the future by such interactions. Since memory cannot be the same as a point in the past, it means that the future cannot be predetermined.

Notes: is this passage a rebuttal to the narrow gap theory? Is this even possible? Is this some kind of spell that foretells the future?

Let's focus on what he said about the third order. The third order of the world rejects the repetition of history, that's right, so in the "Harmony" of the narrow gap, time is limited, the memories of the inhabitants of this place are regressed, so that the history of the continuous cycle is maximized to travel towards a fixed end point, so that the world keeps on doing the same thing over and over again.

Of course, this is not possible, every reincarnation there will be variables, so the burden of managing this gap is given to the dominant person alone, and this can test the dominant person's "Harmony" will is strong or not. It will be a long journey to see whether the long period of mechanized life and management can still keep "Harmony" in the gap.

P128 Original: The five colors of red, blue, white, black and yellow symbolize all things in nature. The directions of east, south, west, north and center represent the seasons of spring, summer, autumn, winter and the equinoxes, as well as the elements of fire, water, wood, gold and earth, which are also the colors of all things.

Notes:

I think it should be "red, purple, white, black, green", it varies from person to person, but if you're going to end with a color, it's purple, the color of the gap.

End

Nov. 30, 2023

He Qiao

Narrow Gap World

Annex 2

Exercises 1

What is the gap?

Gaps can be categorized into narrow gaps and broad gaps, with the broad gap being a large gap consisting of a single, or multiple gaps controlled by multiple dominants, a team, or a group of people. The broad gap is actually now realized Think about what is around us that can be considered a broad gap?

Clever students have already guessed, yes, it's cell phones.

The cell phone is the product of the materialization of a broad interstitial. That's because it's one large gap controlled by multiple dominants.

For a cell phone, the person who uses it is also one of the dominants , but he has minimal control over it, and even if the user has control over it, it is only momentary, that is to say, the user can only control something that is constantly changing and is a variable.

In that case, the burden on the user will be much heavier and the pressure will naturally be high.

As for the narrow gap, since the dominant is the user himself, he is free to control and change the corresponding things, and the pressure is much less.

With a cell phone, there are many software programs that you use regularly. Let's take an example of a program, of course, one that is connected to the Internet. (In this day and age, there are still programs that don't use the Internet, but they are usually not popular.) This software, which is connected to the Internet, must be operated by multiple staff members in the background, and you're just a visitor who came and went in it. You can visit as often as you like, but you will not be the dominant of this software, which is what I consider to be a broad gap.

As for the narrow gap, you can naturally associate it with a computer, and while there will still be a lot of networked software, at least, you can control something a little bit more, can't you?

Yes, that's it, explore with that in mind, what you have to do is to keep expanding your dominance in this gap so that you have less pressure to use it, but how are you going to do that?

For cell phones, just a video watching app for example, you definitely have to charge a VIP membership to be able to watch more content, right? It's a way to enhance dominance, except that the membership will expire. (laughs)

Whereas with computers, if you want to improve dominance, there are many ways to do it, or at least the range of options available to you is expanded and the pressure to dominate the gap is reduced.

Using the four computers "Wild, Harmony, Lucky, and Novelty" as a complete narrow gap is, in my opinion, the best way to increase dominance, as I have separated each of the computer's functions: "Harmony" represents the mechanized process of using the software; "Wild" represents a single rule that seeks excitement and constantly challenges the stronger; "Novelty" represents the constant discovery of new things; "Lucky" represents things that are dangerous but can fertilize the human heart. These four things, I believe, are the most basic things that a person needs on a daily basis.

Like this, using four computers to break down its functions can drastically increase the user's dominance and reduce stress. Okay, that's all for this one.

Exercises 2

What is the point of studying the gap?

First of all, the concept of gap is definitely not a pipe dream, whether it is a narrow gap or a broad gap, it has practical applications. If you research it in advance and pass it on to the next generation, and the next generation explores it on their own, the effect is definitely different.

I don't think there's much research value in broad gaps because cell phones nowadays already implement broad gaps brilliantly.

Now it is the narrow gap that needs to be realized further, because the narrow gap is a large gap that is controlled by you.

Exercises 3

What do you hope real life will become?

I'd like the phone to be used only for scanning or swiping codes, making phone calls, like a pass, without much utility, but of course that's not going to happen.

Anyway, reduce the entertainment, exploration, and social functions of your phone, reduce the functionality of your phone, and use your computer more for those extra functions.

Exercises 4

Evaluate the book Narrow Gap World

The book Narrow Gap World is a theoretical book that tends to be enlightening, without being too theoretical, the whole text is easy to understand, but never simple. I just want to use this book to spread the word about the new concept of gap world, because no one has ever noticed or recognized it. Due to my personal limitations, I didn't delve into some of the concepts as it would require a significant investment of time and effort, so please bear with me.

It is my hope that this book will lead more and more people to discover the field of gaps world, to study them in greater depth, and to lead a clear path for future generations in this age of constant materialization of gaps.

Printed in the USA
CPSIA information can be obtained
at www.ICGtesting.com
LVHW051823110124
768548LV00071B/2902